At the Pond

by Carol Pugliano-Martin

Look at the deer.
The deer is running.

Look at the fish.

The fish is swimming.

Look at the frog.

The frog is hopping.

Look at the bird.
The bird is flying.

Look at the duck.
The duck is walking.

Look at the squirrel.
The squirrel is climbing.

Look at the spider.

The spider is spinning.

Look at the boy.
The boy is fishing.